WHAT'S INSIDE?

MY BODY

DETROIT PUBLIC LIBRARY

CHILDREN'S LIBRARY
5201 Woodward
Detroit, MI 48202

DATE DUE

AEU1938-13

CL

A DK PUBLISHING BOOK

www.dk.com

Conceived, edited, and designed by DK Direct Limited

Note to parents

What's Inside? My Body is designed to help young children understand the fascinating inner workings of the human body. It shows the most important features like the heart, the lungs and the skeleton, and explains in simple terms what each part does. It is a book for you and your child to read and talk about together, and to enjoy.

Designers Juliette Norsworthy and Sonia Whillock
Typographic Designer Nigel Coath
Editors Simon Bell and Alexandra Parsons
Design Director Ed Day
Editorial Director Jonathan Reed

Illustrator Richard Manning
Photographers Steve Gorton and Paul Bricknell
Writer Angela Royston

First American Edition, 1991

2 4 6 8 10 9 7 5 3 1

Published in the United States by
DK Publishing, Inc.
95 Madison Avenue
New York, New York 10016

Copyright © 1991 Dorling Kindersley Limited, London
First paperback edition, 1999

All rights reserved under International and Pan-American Copyright Conventions. No part of this publication may be reproduced, stored in a retrieval system, or transmitted in any form or by any means, electronic, mechanical, photocopying, recording or otherwise, without the prior written permission of the copyright owner.
Published in Great Britain by Dorling Kindersley Limited.

Library of Congress Catalog Card Number: 91-60535

ISBN: 1-879431-07-6 (hardcover)
0-7894-42930 (paper)

Printed in Italy

WHAT'S INSIDE?

MY BODY

DK PUBLISHING, INC.

UNDER MY SKIN

I know how my body looks on the outside, but I wonder what is inside? My skin and flesh are soft and under them I can feel hard, knobby bones.

My hair grows longest and thickest on my head.

Skin covers my body like a stretchy suit. It stops germs from getting inside.

I can feel bones all over my body, except here on my stomach.

Hundreds of small, fine hairs grow all over my skin. They help to keep me warm.

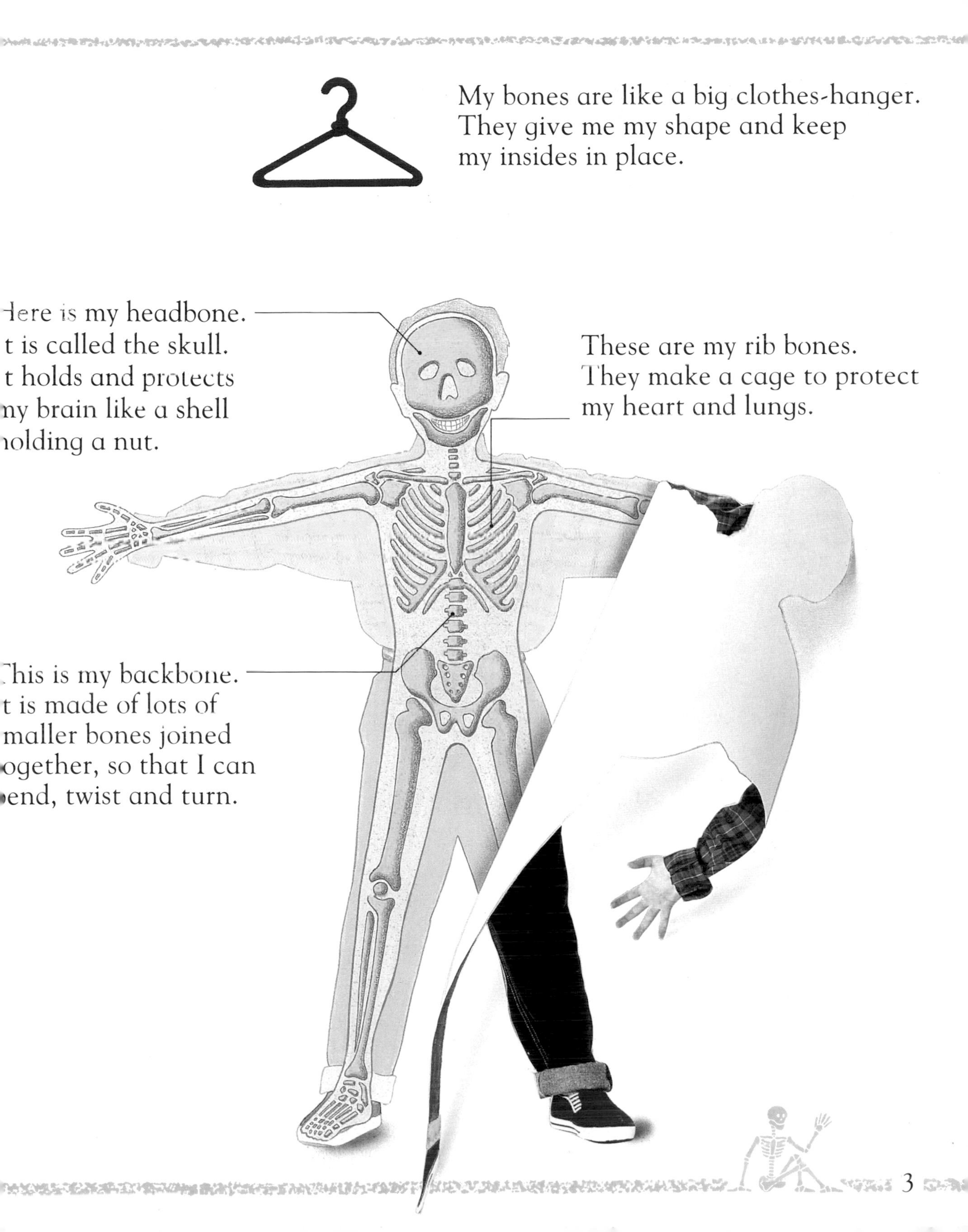

My bones are like a big clothes-hanger. They give me my shape and keep my insides in place.

Here is my headbone. t is called the skull. t holds and protects ny brain like a shell holding a nut.

These are my rib bones. They make a cage to protect my heart and lungs.

This is my backbone. t is made of lots of maller bones joined ogether, so that I can bend, twist and turn.

MY HEAD

My head is a very important part of me because it holds my brain. I think, feel and remember with my brain. My eyes, ears, mouth, nose and skin tell my brain what is happening in the world outside.

I smell and breathe throu the two holes in my nose. These are called nostril

My lips and tongue hel me to talk and to taste different flavors.

My neck holds up my head. The muscles in my neck are very strong.

My brain is like a computer. It is my body's control center. Messages are sent between my brain and all the different parts of my body.

My nose tells my brain what I am smelling.

This is my windpipe: it takes air to my lungs.

/Iy tongue tastes ny food and moves it ıround my mouth ıs I chew.

This is my foodpipe: it goes down to my stomach.

MY EYES

My eyes are like television cameras. They send moving pictures of the world outside back to my brain, which then works out what I can see.

My eyelashes help to stop dust from getting in my eye.

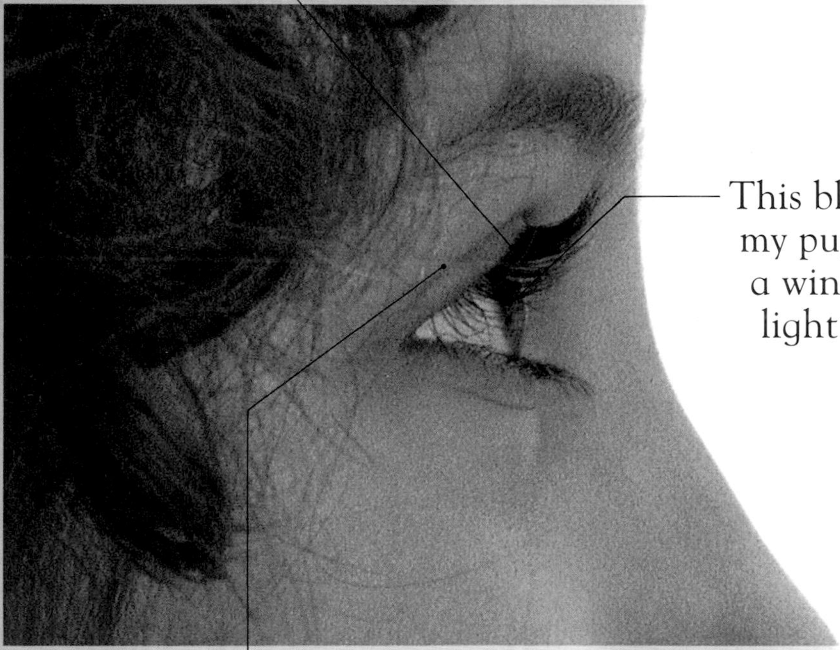

This black circle is my pupil. It is really a window that lets light into my eye.

My eyelids are like curtains. They cover my eyes when I sleep and when I blink. I blink about 15 times a minute.

This is my eyeball.
Inside is a kind of clear jelly.
It keeps my eyeball round and smooth,
and light can pass through it.

Like a camera,
my eye has a lens
to help me to see
.early and sharply.

'ears drain through
ıese special tubes.
'ears keep my eyes clean.
ut nobody knows why we cry
:ars when we are unhappy.

This nerve
carries the picture
from my eye back
to my brain.

MY EARS

All I can see are the outer ears on the sides of my head. Most of my ear is out of sight inside my head. This hidden part of my ear sends messages to my brain, telling me what I can hear.

The outer part of my ear is like a funnel. It directs sounds into my ear passage.

There are no bones in this part of my ear. I can bend it like a piece of rubber.

I have one ear on either side of my head. Two ears help me to tell exactly where sounds are coming from.

This is my eardrum. When a sound hits this piece of skin, it vibrates just like a drum.

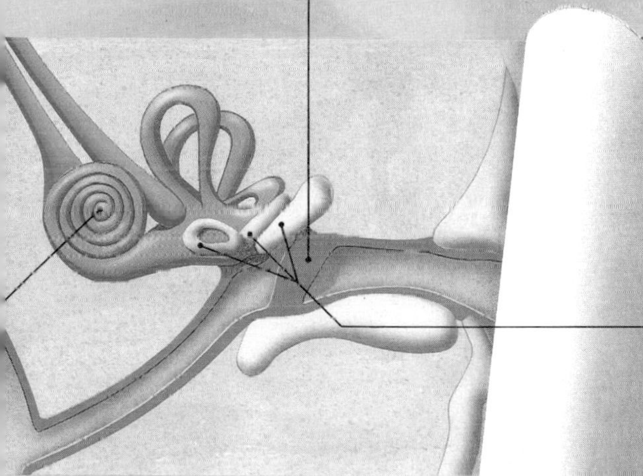

These three small bones carry the vibration through to my inner ear.

This is my inner ear. The nerves in here tell my brain what I am hearing.

MY CHEST

When I run very fast I can feel a thumping in my chest. This is my heart beating. Sometimes I get really out of breath and then I breathe in all the air I can.

These are my nipples. Mothers use them to feed their babies with milk. Children and men do not use their nipples.

When I put my hand here I can feel my heart beating.

When I put my hand here, I can see and feel my chest going in and out as I breathe.

breathe in air through my
ose and mouth. The air
omes down my windpipe
nd fills up my lungs.

This is my heart.
It keeps my blood
moving all around
my body. My blood
never stays still; it is
always moving.

Iere are my
ungs. They
ake in oxygen
rom the air.
need oxygen
o keep me alive.

This big flat
muscle is my
diaphragm. It
helps me breathe.

MY STOMACH

When I am hungry my stomach feels empty and I can hear it rumbling. When I have eaten a lot it feels stretched and full.

Put your ear against a friend's stomach. Can you hear the juices inside making gurgling sounds?

I can pinch the flesh on my stomach. The fat is like a padded jacket that keeps me warm.

Before I was born I was joined to my mother by a special cord. After I was born it was tied and cut, leaving me with a belly button.

The food that I swallow goes on a long journey through my body. On the way, all the good things that my body needs are passed through to my blood.

This is my foodpipe. It carries food from my mouth down into my stomach.

In my stomach special juices break the food into small pieces.

y food then
ısses through
is long pipe,
ıere it is
ashed into
gooey soup.

The bits of food my body does not want go right to the end of the tube. They leave my body when I go to the bathroom.

MY LEGS AND ARMS

I use my legs and feet to walk, run and jump.
I use my arms to push, pull and carry.

My elbow is where my arm bends and straightens.

My knee is where my leg bends. Is your knee knobby or smooth?

Exercise helps to keep my muscles supple and strong.

My ankle lets me wiggle and twist my foot from side to side.

My toes help me to balance and bend my foot as I walk.

Muscles help me move my bones. I have muscles all over my body. They are joined to my bones by long stretchy bands called tendons.

I have muscles to move my hand...

...muscles to bend and straighten my arm...

...big thigh and calf muscles to move my leg...

...and muscles to move my foot.

When I bend my arm, the muscle bunches up. When I stretch it, the muscle flattens out.

MY HANDS

I can wave, draw and pick up all sorts of things with my hands. I use my hands to feel things too. The skin on my fingertips can feel even the tiniest grain of sand.

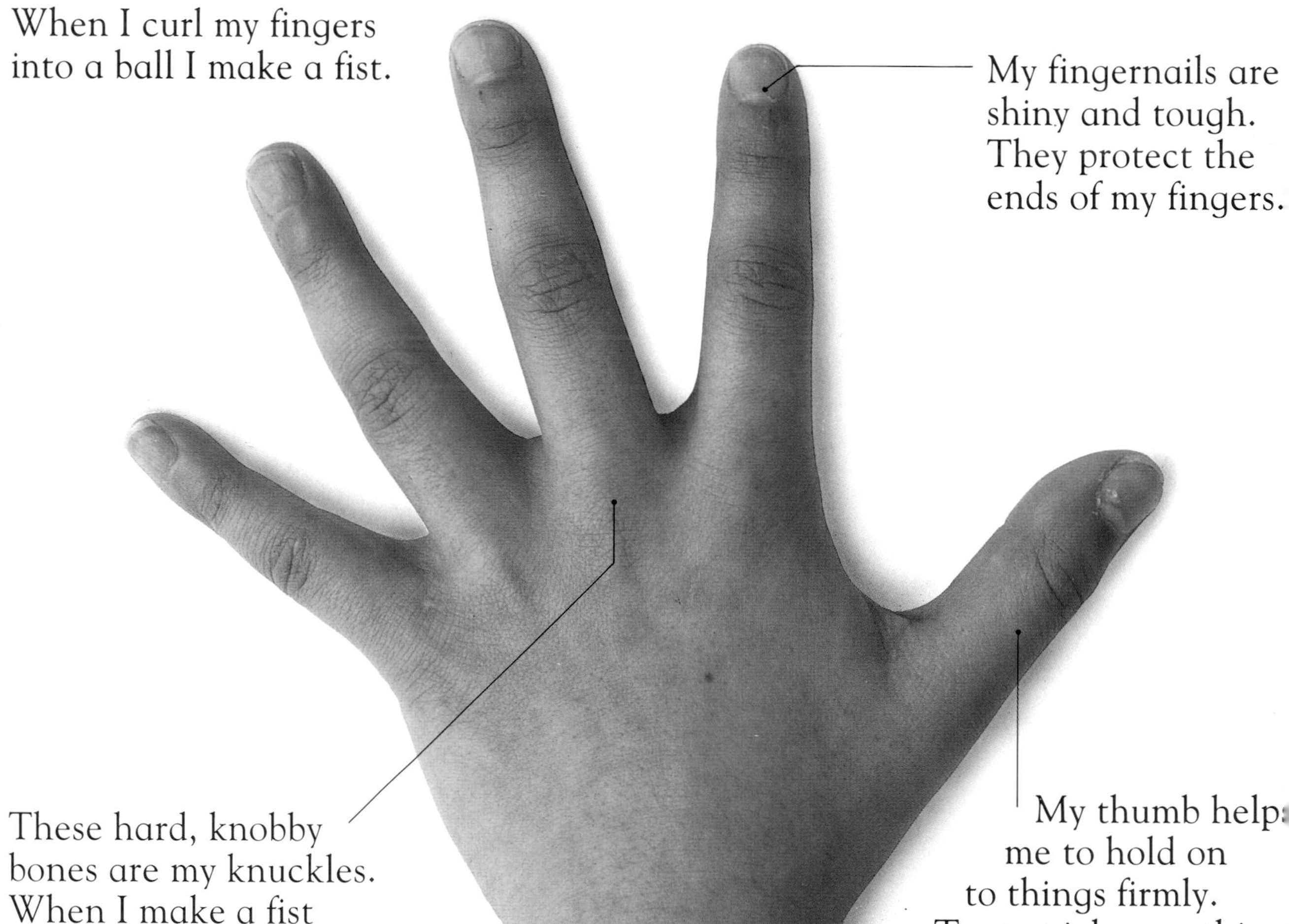

The nerves in my fingertips feel all sorts of things – hot and cold, hard and soft, rough and smooth.

ly fingers are orked by tendons, hich are like pieces f string attached ɔ my fingerbones.

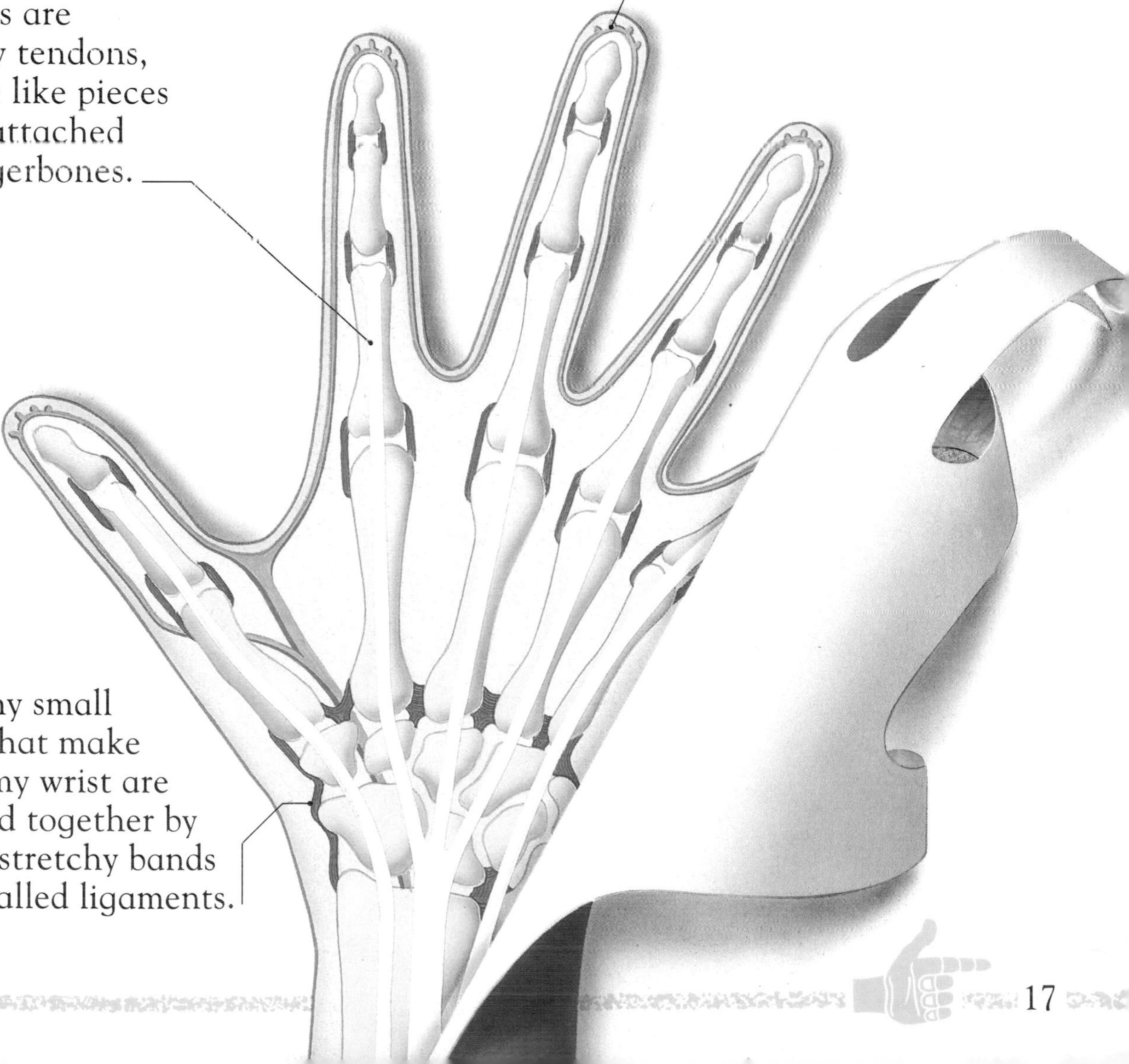

The many small bones that make up my wrist are held together by stretchy bands called ligaments.